Gisvani L. Vasconcelos
Déborah M. Freitas
Josiana S. C. Siebra

In vitro production of bovine embryos

Gisvani L. Vasconcelos
Déborah M. Freitas
Josiana S. C. Siebra

In vitro production of bovine embryos

A systematic review

ScienciaScripts

Imprint

Cover image: www.ingimage.com

This book is a translation from the original published under ISBN 978-620-2-40746-5.

Publisher:
Sciencia Scripts
is a trademark of
Dodo Books Indian Ocean Ltd. and OmniScriptum S.R.L publishing group

120 High Road, East Finchley, London, N2 9ED, United Kingdom
Str. Armeneasca 28/1, office 1, Chisinau MD-2012, Republic of Moldova, Europe
Printed at: see last page
ISBN: 978-620-8-32409-4

PRESENTATION

This book is part of the results achieved by the authors Deborah Marinho Freitas and Josiana Sampaio do Carmo Siebra, in the development of their Conclusion Monographs for the Specialization Course in Applied Biochemistry and Molecular Biology, with an Emphasis on Health, Environment and Agriculture, at the Vale do Acarau State University - UVA.

This paper is a literature review that was designed to take into account the importance of understanding the process of *in vitro* oocyte maturation (IVM) and *in vitro* fertilization (IVF) in order to improve the process of bovine *in vitro* embryo production (BIP). Many studies have been carried out to evaluate the stages of IVM and IVF in order to obtain better results, adjusting all the variables involved in these techniques in the best possible way. In this way, an accurate understanding of the metabolic needs of the oocyte/embryo during IVP is essential in order to establish the ideal condition and provide the competence of the greatest possible number of embryos produced *in vitro.*

As a result, the production of materials that compile the main studies on the PIVE technique is favorable for improving *in vitro* culture conditions, which could contribute to a surprising increase in the number of embryos generated in Brazil.

SUMMARY

The *in vitro* production of embryos (IVP) is a technique used in the reproduction of animals and has become a tool of continuous growth in Brazil, since it is a technique responsible for the rapid genetic development of production animals, a fact that has allowed this technique to spread from research to commercial application. IVP comprises the stages of *in vitro* maturation (IVM) of the oocytes, classification of the sperm and *in vitro* fertilization (IVF) with subsequent *in vitro* embryo culture (IVEC) and development up to the blastocyst stage. For the technique to be successful, all the stages must be carried out correctly, since one interferes with the success of the next. In order to improve the low efficiency (35 to 40% of blastocysts) in bovine IVP, it is necessary to understand the IVM stage. A better understanding of this stage makes it easier to understand the changes and needs during the growth and development of the oocyte. It is also essential to intensify the production of a large number of embryos during the IVC. IVM is a technique that consists of obtaining mature oocytes under artificial conditions and comprises three stages: nuclear maturation, cytoplasmic maturation and molecular maturation. In order for the oocytes to develop properly until they are ready to be fertilized, these processes must occur in perfect synchrony. This literature review aims to describe the mechanisms of the IVP technique and its use in intensifying the use of females of high zootechnical value, as well as demonstrating the importance of the IVM technique, highlighting nuclear, cytoplasmic and molecular events.

Keywords: Cattle. Cultivation. Embryos. Oocyte. PIVE.

SUMMARY

1 INTRODUCTION

The development of breeding biotechniques such as artificial insemination (AI) and IVP is important for the genetic improvement of herds. IVP has been used with the aim of improving commercial use, since one of the main impasses of the technique is the variability *in the* results of *in vitro* embryo development (DAYAN; WATANABE; WATANABE, 2000).

In vitro matured oocytes, when compared to *in vivo* ones, have lower blastocyst rates after fertilization and CIVE (TAKAGI *etal.,* 2001). It can therefore be assumed that maturation is still a problem for IVF. Thus, the success of oocyte maturation can be affected by many factors, including cumulus oocyte complex (CCO) morphology, oocyte competence and maturation conditions (GONQALVES *et al.,* 2002).

After the oocytes have been retrieved (obtained), three biological steps that take place *in vivo* are carried out *in* the laboratory for IVP to take place: MIV, FIV and CIVE (GARCIA; AVELINO; VANTINI, 2004). These techniques comprise a series of stages, but MIV is certainly one of the crucial ones, as it is during this period that the oocyte acquires the capacity to go on to the next events.

According to Warssaman and Albertini (1994, p. 79-122) cited by Santos (2010, p.25), maturation represents the final stage in the preparation of the oocyte for fertilization, and therefore the structural and biochemical changes that occur during this period are essential for this process to be completed. However, for this to happen, the oocyte needs to have meiotic competence, which is gradually achieved during the final phase of oocyte growth.

Despite the progress made in the last 20 years with IVP in cattle, the rates remain stable, i.e. the rate of nuclear maturation and fertilization is around 80% and the rate of blastocyst formation is between 35-40% (PONTES *et al.,* 2011), since only around 30% (PONTES *et al.,* 2009) of these embryos are capable of producing a pregnancy after transfer. However, when the oocytes are matured *in vivo,* the blastocyst rates are remarkable when compared to those of oocytes matured *in vitro.* These results show that maturation is still one of the limiting steps in advancing the efficiency of assisted reproduction techniques.

Although the embryo transfer technique is widespread worldwide, the variability of the response to treatment is still limited. As a result of this limitation, PIVE has expanded in recent years as an alternative technique to increase the number of embryos obtained, since with PIVE it is possible to produce embryos regardless of the stage of the donor's estrous cycle and the process can be repeated without negatively interfering with the number of oocytes recovered (ANDRADE *et al.,* 2012). Therefore, the aim of this work is to describe the procedures of the IVP technique and its use in intensifying the use of females of high zootechnical value, as well as demonstrating the importance of the IVM technique, highlighting nuclear, cytoplasmic and molecular events.

2 CHAPTER 1 - OBJECTIVES

2.1 General

- Understand the ideal mechanisms for successful *in vitro* production of bovine embryos.

2.2 Specifics

- To highlight the changes *in the in vitro* maturation of oocytes and their essential characteristics for embryonic development;
- Emphasize the cytoplasmic, molecular and nuclear changes that occur during *in vitro* maturation *in* the acquisition of oocyte competence;
- Present the importance of the stages during the *in vitro* production of bovine embryos.

3 CHAPTER 2 - LITERATURE REVIEW

3.1 Oogenesis

Oogenesis is the process of development of the female gamete and begins during the intrauterine stage of development (see figure 1). The primordial germ cells originate from the endoderm and move to the genital ridges to colonize the future gonads. During this phase of migration and colonization, intense mitotic divisions take place, which are necessary for the formation of oocyte "stocks". After arriving in the gonads in formation, these cells stop dividing mitotically and begin meiotic division (VAN DEN HURK; ZHAO, 2005).

Figure 1. Schematic drawing of the mammalian ovary.

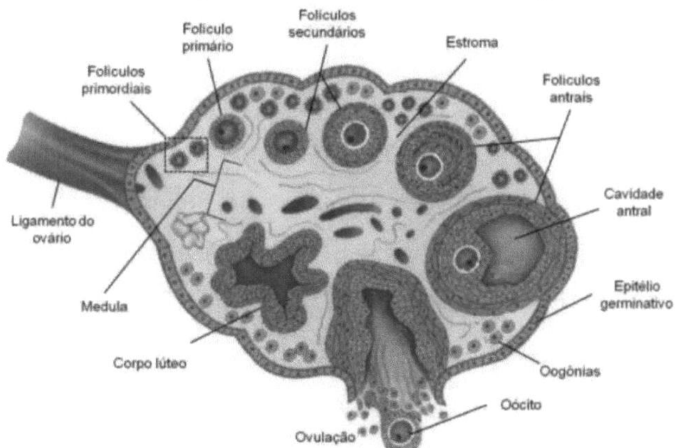

The mammalian ovary contains follicles at different stages of development (primordial, primary, secondary, tertiary and pre-ovulatory), corpus luteum and oogonia on the ovarian surface (TILLY;

JOHNSON, 2007). Source: Adapted from http://academic.rcc.edu/moore/docs/Bio30/Lecture%208%20-%20Oogenesis. pdf.

In cattle, the process of meiotic division begins at 75-80 days of fetal life, but the cell cycle is blocked in prophase I, at the diplotene stage. The oocytes remain quiescent until the pubertal stage, when shortly before ovulation, under the stimulus of luteinizing hormone (LH), they resume meiosis and the cell cycle progresses from prophase I to metaphase II. *In vitro,* the resumption of meiosis is stimulated by the simple removal of the oocyte from the follicular environment (SIRARD *et al.,* 2006).

The follicular environment is responsible for maintaining the oocyte in prophase I during folliculogenesis and is also a determining factor in the resumption of meiosis during maturation. The oocyte resumes meiosis after the induction of the LH surge, when the germinal vesicle (GV) breaks down and the first polar corpuscle is extruded (AKTAS *et al.,* 1995; TSAFRIRI *et al.,* 1996; BILODEAU-GOESEELS, 2011).

Biochemical, nuclear and cytoplasmic events begin with the formation of the primordial follicle and continue until the moment of ovulation with oocyte competence (GANDOLFI; GANDOLFI, 2001). This competence is progressively acquired during the final stages of folliculogenesis, through various cellular and molecular changes that give the oocyte the ability to complete meiotic division and continue its development (COTICCHIO *et al.,* 2004), since follicle and oocyte development are parallel and functionally related events (AGUILLAR *et al.,* 2001).

3.2 Principles of oocyte maturation

Age, puberty, reproductive conditions, sward and genetic makeup are important characteristics that will influence the productivity and increased reproductive life of the animal (CARDOSO; NOGUEIRA, 2007).

The animal's reproductive competence is one of the variables with the greatest economic impact on the entire production process, and the animal always needs an environment of thermal comfort in order to maximize its efficiency. In this way, reproduction is characterized as the most important promising factor in cattle breeding, directly affecting the productivity levels of a herd (RADOSTITS, 2001).

In this sense, the capacity for oocyte development can be influenced by the physiological conditions of the donor, such as the stage of the estrous cycle, the stage of follicular development, among others (LONERGAN; FAIR, 2008). Although many factors are unknown, it should be considered that oocyte development within the follicle is controlled by endocrine and local factors, such as hormones, growth factors and peptides. Thus, the conditions in which the oocyte's nuclear and cytoplasmic maturation takes place certainly play a fundamental role in the acquisition of competence, which is essential for embryonic development. Therefore, in order to establish a culture system that enables the production of a greater number of good quality embryos, it is necessary to understand the mechanisms involved in oocyte maturation (GOTTARDI; MINGOTI, 2009).

During IVM, oocytes undergo a number of changes at the molecular and structural level to ensure that the process is completed optimally. *In vitro,* the resumption of meiosis occurs abruptly and spontaneously, regardless of acquired competence. Thus, some oocytes resume meiosis without acquiring full competence (ADONA; LEAL, 2006; GILCHRST; THOMPSON, 2007), i.e. they have not yet completed the cytoplasmic machinery to support development (GILCHRIST; THOMPSON, 2007). Inadequate maturation of either the nucleus or the cytoplasm makes fertilization impossible and increases the occurrence of polyspermy, parthenogenesis and blocked embryonic development. With the completion of the nuclear and cytoplasmic maturation processes, the oocyte becomes suitable for successful fertilization and early embryonic development (MINGOTI, 2005).

In view of this, understanding the mechanisms involved in oocyte maturation is essential for implementing a cultivation system that enables the production of a greater number of good quality embryos (VAN DEN HURK; ZHAO, 2005; GOTTARDI; MINGOTI, 2009).

3.3 *In vitro* oocyte maturation

IVM is a technique that consists of obtaining mature oocytes under artificial conditions to prepare them for fertilization and subsequent embryonic development (figure 2) (GONQALVES *et al.,* 2008). It is a stage in which the oocyte needs to complete its development in order to receive sperm under artificial conditions.

The vast majority of oocytes selected for IVF have a low capacity for embryonic development. In live animals, the oocytes acquire embryonic capacity during follicular development, where the oocytes undergo molecular changes, allowing them to pass through the initial stages of development without major losses. In live animals, maturation takes between 18 and 24 hours, with 5% CO2 and saturated humidity, at a temperature of 38.7°C. *In vitro*, maturation takes place using culture media such as TCM 199 (GONQALVES, 2007). This medium is supplemented with buffers, amino acids, luteinizing hormone (LH), follicle stimulating hormone (FSH) and estradiol, fetal bovine serum (FBS) and bovine serum albumin (BSA), among others (MAINARDES, 2010).

The technique selects oocytes with superior quality (Grades I, II and III) that have the capacity to develop to later stages. In this way, oocytes are classified according to the number of cumulus cell layers and intracellular quality into: Grade I - compact cumulus, containing more than three cell layers and homogeneous cell cytoplasm; Grade II - compact cumulus, partially present around the oocyte, with less than three cell layers. The cytoplasm of the cell has unevenly distributed granules; Grade III - cumulus present, but expanded. The cytoplasm is contracted, degenerated, fragmented or has vacuoles. After the oocytes have been selected, they are inserted into a medium for maturation (GONQALVES *et al.,* 2008).

Figure 2. Hoechst staining to assess the maturation and cell number of bovine embryos.

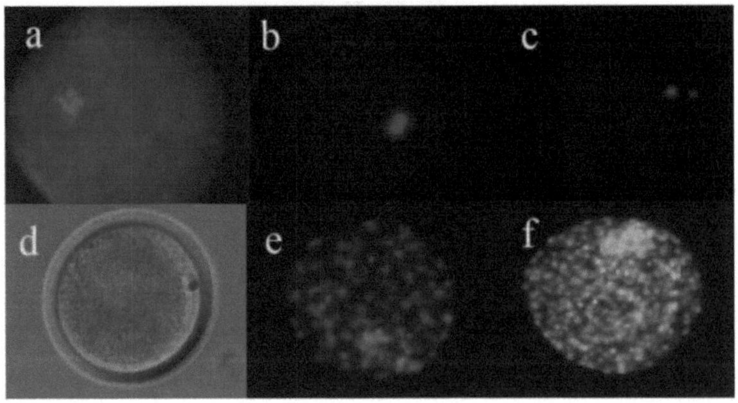

Oocyte in germinal vesicle (a), intermediate stage (b), metaphase II (c), release of 1st polar corpuscle (d), embryo with ±110 cells (e) and embryo with ±360 cells (f). Source: Guemra *et al.,* 2013.

During oocyte maturation, endogenous hormones and promoting factors are secreted which stimulate the synthesis of hyaluronic acid by cumulus cells, inducing their expansion (GONQALVES *et al.,* 2002). These cells play a key role in the first hours of maturation, protecting the oocyte from oxidative stress (MERTON *et al.,* 2003). Consequences of this stress can cause negative alterations in the maturation and fertilization processes of the oocytes, as well as in the culture of the probable embryos. To reverse this situation, it is necessary to reduce the production of ROS or increase the amount of antioxidants available (ANDRADE *et al.,* 2010).

In order to understand the importance of the quality of oocyte maturation for subsequent development to the embryonic stage, it is essential to clearly specify the different forms of competence assumed by the oocyte during the maturation process. These forms can basically occur in three different processes: molecular, cytoplasmic and nuclear events (GOTTARDI; MINGOTI, 2009).

IVM is characterized by the progression of development from the germinal vesicle stage to metaphase II (MII). From there, the oocyte is prepared to be fertilized and undergo embryonic development. Changes in the oocyte occur in both the nucleus and the cytoplasm, but independently, so that neither interferes with the other, thus ensuring productive cell development (FERREIRA, 2010). In this sense, during the IVM process, the oocyte undergoes nuclear and cytoplasmic changes which will be discussed below.

3.3.1 Cytoplasmic maturation

Cytoplasmic maturation can be conceptualized as a set of processes by which the mammalian oocyte becomes a cell capable of fertilizing and supporting early embryonic development (ANGUITA *et al.,* 2007).

Cytoplasmic events include the breakdown and synthesis of

proteins and molecular modifications (KUBELKA *et al.*, 2000), redistribution of intracellular organelles (STOJKOVIC *et al.*, 2001) and maturation of calcium ion release mechanisms (Ca^{2+}) (WANG *et al.*, 2003). Cytoplasmic maturation comprises a series of biochemical events that enable the oocyte to expose its developmental potential.

In this sense, with the reorganization of organelles (FERNANDES, 2004), there are changes in the Golgi complex, migration of mitochondria to a more central region and cortical granules to a peripheral region (HUMBLOT *et al.*, 2005; SIRARD *et al.*, 2006), storage of proteins and RNA (ANGUITA *et al.*, 2007) and parthenogenetic cleavage (CARNEIRO *et al.*, 2001). The most common cytoplasmic changes are an increase in lipid storage (CHAVES, 2010), among others. The increase in lipid stores probably occurs so that the oocyte has enough energy to complete its development until the blastocyst. An oocyte that does not complete its cytoplasmic maturation becomes unviable for subsequent IVF (CHAVES, 2010).

The complex events that take place during oocyte maturation include not only the correct dynamics of chromosome separation during nuclear maturation, but also the storage of energy and other indispensable factors for the process of maturation, fertilization and embryonic development (SANDRI, 2007).

The improvement of the *in vitro* cytoplasmic maturation technique can be achieved by improving the embryonic development capacity of oocytes from small antral follicles.

Growth factors produced by the cells during follicular development, such as: insulin-like growth factor 1 (IGF-1), epidermal growth factor (EGF) and insulin-like growth factor (IGF). fibroblast growth factor (FGF), growth and differentiation factor 9 (GDF-9) and bone morphogenetic protein 15 (BMP-15) have already been tested during IVM and have improved oocyte capacitation (GONQALVES *et al.*, 2007).

Despite the large amount of information available on the subject, no ideal method has yet been found for assessing cytoplasmic maturation other than fertilization itself and the subsequent normal development of the embryo and fetus (KRISHER, 2004).

3.3.2 Molecular maturation

Molecular maturation corresponds to the growth and maturation phases of the oocyte, and is defined as the transcription, storage and processing of the mRNA expressed by the chromosomes (FERREIRA *et al.,* 2008), which will then be translated into proteins by the ribosomes.

The proteins derived from these RNAms are involved both in maturation and in subsequent cellular events: fertilization, pro-nucleus formation and early embryogenesis, and must therefore be stored until they are used (SIRARD, 2001). Transcription, the storage of mRNA and proteins occurs during follicular growth and ceases when the germinal vesicle breaks down (GVDB) with the resumption of meiosis, once the chromosomes become condensed and inactive (GANDOLFI; GANDOLFI, 2001).

The oocyte is primarily responsible for coordinating the RNAms and proteins that were accumulated during its growth phase (FAIR *et al.,* 2007; HAMATANI, *et al.,* 2008). The transcriptional profile of bovine oocytes matured *in vitro* and *in vivo* has shown distinct pathways and variations in the number of genes, which can alter embryonic development after fertilization (KATZ-JAFFE *et al.,* 2009; MAMO *et al.,* 2011).

3.3.3 Nuclear maturation

During IVM, oocytes undergo various nuclear and cytoplasmic changes. Nuclear events include the breakdown of the germinal vesicle, disappearance of the nucleolus, chromatin condensation, extrusion of the first polar corpuscle and formation of the second meiotic spindle (MEINECKE *et al.,* 2001). The chromosomes then condense and are arranged in the central plane of the metaphase axis, before being compacted and the first polar corpuscle is expelled, characterizing the evolution from metaphase I to MII (FERREIRA, 2010). It consists of a series of transformations that take place in the nucleus, which leads to the restart of meiosis from the VG stage to the metaphase of the second meiotic division.

Nuclear maturation refers to the resumption of the first meiotic block with the breakdown of the germinal vesicle and progression to MII (MINGOTI, 2000). It is the cascade of nuclear events that occur when the oocyte is stimulated by pre-ovulatory

LH, or by the removal of the oocyte from its follicular environment (SIRARD, 2001).

Nuclear maturation lasts approximately 22 to 24 hours in cows and involves the breakdown of the nuclear envelope and the progression of meiosis (LUCIANO *et al.,* 2009). Under the influence of hormones
gonadotrophic, the oocyte resumes the cell cycle from the diplotene phase of prophase I, passes through the stages of metaphase I, anaphase I, telophase I (end of the first meiotic division) and progresses to the metaphase stage of the second meiotic division (MEINECKE *et al.,* 2001; VAN DE HURK; ZHAO, 2005). In the interval between prophase stages I and MII, the chromosomes condense and the nuclear envelope breaks down (GVBD), marking the beginning of nuclear maturation (MEINECKE *et al.,* 2001; JONES, 2004).

The nuclear maturation of pre-ovulatory oocytes is organized by components of the follicular fluid, the interaction of oocytes and follicular cells with hormones. As for pre-antral and antral oocytes, the vesicle breaks down when they reach the ideal diameter according to the species (CHAVES, 2010).

The homologous chromosomes are divided into two groups, with half of the original number of chromosomes remaining in the oocyte (haploid cell) and the other half being included in the first polar corpuscle. At the end of the first meiotic division, the cytoplasm is divided asymmetrically (CAN *et al.,* 2003), generating two cells of different sizes: a small one called the polar corpuscle and a larger one, the secondary oocyte. After nuclear maturation, the oocyte remains in this stage of the cell cycle (MII) until fertilization (MAYES; SIRARD, 2001).

At the end of meiotic division, the cytoplasm divides irregularly to form a smaller polar corpuscle and a larger secondary oocyte. At the end of nuclear maturation, the oocyte continues in the MII stage until fertilization (GOTTARDI, 2009).

3.4 Media used in IVM of bovine oocytes

A wide variety of media have been used for IVM of oocytes, however, the most widely used base medium is *Tissue Culture Medium* (TCM 199) (VARAGO *et al.,* 2008). Generally, some supplements are added to this medium, such as SFB, L-glutamine, sodium bicarbonate, HEPES, sodium pyruvate, antibiotics and hormones, in order to stimulate *in vitro* development (GUIXUE *et al.,* 2001).

With regard to hormones, it is common to use LH, FSH or a combination of these two gonadotrophins. The addition of gonadotrophins to the IVM medium is intended to stimulate the development of bovine oocytes (GUIXUE *et al.,* 2001). FSH stimulates the production of signaling substances by somatic cells, which causes meiosis to resume, stimulating the expansion of cumulus cells and, consequently, facilitating fertilization (GOTTARDI; MINGOTI, 2009; ULLAH *et al.,* 2008).

There are controversies about the need for these hormones

and the concentrations to be used, but it is known that the addition of gonadotrophins to the oocyte maturation medium increases fertilization capacity and improves subsequent embryonic development (GONQALVES *et al.,* 2002).

Modifications to the media depend on supplementation with an energy source, a protein source or SFB (DODE *etal.,* 2002). The use of serum in maturation media increases the concentration of protein and growth factors, which prevent the zona pellucida from hardening, thus increasing the fertilization capacity of the oocytes, and has been used in the proportion of 10% to 20% of the total volume of the medium (GONQALVES *et al.,* 2002).

Studies have shown that the addition of BSA to the maturation medium increased the expression of receptors for IGF-1 and IGF2R and for IGF- 2 in bovine blastocysts, and the use of SFB in maturation increased the levels of mRNA for HSPA1A, a *heat shock* 70 protein (PARKS *et al.,* 2011). Improved maturation conditions and blastocyst development can be achieved by adding factors to the maturation medium such as linoleic acid (ELSAYED *et al.,* 2006) and leptin (CROSS *et al.,* 2003) and FSH (ULLAH *et al.,* 2008).

The effectiveness of these means is not completely clear, but studies have shown a possible link between the involvement of growth factors in improving oocyte competence (CARNEIRO *et al.,* 2003).

In addition to what has been discussed above, for IVM to be successful it is also necessary for the medium to have the right pH, osmolarity and ionic composition, the temperature of the maturation oven must be ideal (38.5 to 39° C) and with 5% free CO2 and O2, which are important for maturation to take place successfully (GONQALVES *et al.*, 2002).

3.5 *In vitro* fertilization (IVF)

The IVF stage involves the union of oocytes with sperm and depends on the good quality of the sperm involved. The best quality sperm are chosen and separated using the *Swin up* method or the *Percoll* density gradient method. The fertilized oocytes are added to drops of FERT-TALP fertilization medium and incubated at 38.7°C with 5% CO2 and saturated humidity (GOUVEIA, 2011). This medium contains heparin, which is capable of promoting sperm capacitation, epinephrine, which is important for sperm motility, hypotaurine and penicillamine (GONQALVES *et al.*, 2007).

3.6 *In vitro* embryo culture (IVEC)

The bovine IOC refers to the stage of development of the fertilized oocyte up to the blastocyst stage. It is during this period that the separation and subsequent compaction of the blastomeres in the morula stage takes place, followed by the formation of the blastocele, the central region of the blastula (BUENO; BELTRAN, 2008).

These same authors have shown that the necessary requirement for good quality ESCs, and thus good embryo production rates, is the culture conditions. The effects of internal and external elements that can influence embryo metabolism and growth have been analyzed. Temperature and gas atmosphere, composition of the medium, addition of vitamins, amino acids and growth hormones are examples of these elements.

After the fertilization period, the possible developed zygotes are transferred to drops of culture media (CR2), the composition of which depends on the nutritional requirements of each zygote. The drops of culture medium are covered with mineral oil to prevent drying out, and the zygotes are stored for 6 to 8 days until they reach the blastocyst stage (GOUVEIA, 2011).

3.7 *In vitro* embryo production (IVEP)

PIVE is a technique used as an alternative to increase embryo production on a large scale. This technique, combined with artificial insemination and embryo transfer, is responsible for rapid genetic development, especially in production animals. PIVE allows genetically good breeding stock not to be discarded prematurely due to some alteration that prevents natural mating. In this way, quality PIVE depends on good oocyte quality and good sperm quality (DOMINGUES *et al.*, 2011).

IVP comprises the stages of oocyte IVM, sperm sorting and IVF and IVC, where they will develop to the morula or blastocyst stage (VARAGO *et al.*, 2008). The oocytes used in PIVE are obtained from adult females or females that are still entering puberty. It is carried out on young females, females that do not show oestrus or genetically superior females. These animals may or may not undergo hormonal treatment, which aims to stimulate follicular growth and obtain the highest possible number of oocytes per ovary. This hormonal application may vary according to the age of the animal (BRAGANQA, 2007). On the other hand, ovaries can be obtained from slaughterhouses without hormonal stimulation (BAGGIO, 2008).

In this way, ovaries are taken to the laboratory in phosphate-buffered saline (PBS) or 0.9% sodium chloride (NaCl) solution (LOPES, 2008). The oocytes can be collected by dissecting the follicles or by aspirating them with needles attached to syringes. Follicle dissection allows a large number of oocytes to be obtained (an average of 20-40 per ovary), but it is a time-consuming process and requires good experience on the part of the person handling it. Follicular aspiration, on the other hand, is a simple and quick process and allows better quality oocytes to be obtained because they have a large number of cumulus cells (CROCOMO *et al.,* 2012).

3.7.1 History of biotechnology

The first bovine was born through IVF in 1982 in the United States (RUBIN *et al.,* 2009).
follicular aspiration, IVM, IVF, IVC and subsequent embryo

transfer. IVF is becoming an auxiliary instrument in animal breeding, being a tool in the formation of databases and in the preservation of endangered animals (DOMINGUES *et al.,* 2011). IVF is indispensable for new techniques such as nucleus transfer, cloning and gene manipulation (GONQALVES *et al.,* 2007).

Commercial bovine IVF in Brazil began in 1998 and has increased considerably over the years. It is essential to understand the improved biological mechanisms involved in the process of IVF of embryos obtained from the oocytes of slaughtered cows. These oocytes can be obtained from live animals using the ultrasound-guided follicular aspiration (OPU) technique, which makes IVF possible. During IVC, the embryo goes through the cleavage stages until it reaches the blastocyst stage. At this stage, they are introduced into the recipients, with the aim of making genetic use of the animal (BUENO; BELTRAN, 2008). To obtain good results in IVF, it is important to have the lowest pH variation in the culture media, with the ideal pH being between 7.3 and 7.5 (LOIOLA, 2013).

It is of the utmost importance that material losses during the IVF process are low, due to the high financial investment, so a number of steps must be taken, both in the field and in the laboratory. Hormonal evaluation and management of donors and recipients, choice of semen, early identification of pregnancy, embryonic losses, fetal sexing using ultrasound and evaluation of the laboratory's quality control are all criteria that should be analyzed (GOUVEIA, 2011).

The advantages of IVF include an increase in IVFP in cattle, in addition to improving genetic gain in meat or milk production, the creation of frozen oocyte banks, the gestation of immature females (SOUZA, 2007), as well as determining and controlling the sex of the animal, ease of importing and exporting genetic material to databases, among others. However, there are limitations associated with the processes involved in IVF. These limitations are related to the lack of efficiency in producing blastocysts from oocytes submitted to IVM and the low variability of the process, as well as the fact that embryos produced *in vitro* are less viable and less resistant to freezing (MONTAGNER *et al.*, 2000).

3.7.2 Growing media for PIVE

Culture media are essential for a successful IVP, as they directly interfere with embryonic development. Alterations in the preparation of the medium, due to a worn-out reagent or human error, can cause damage.

Researchers interested in the study of PIVE started the first experiments with the aim of establishing a protocol that would allow embryos to be handled, and one of the first obstacles was the nutritional and physico-chemical limitations present in the culture media. At the beginning of the 20th century, better quality media was produced, which led to successful embryo collection and CIVE (ANDREOTI, 2007).

The milestone in the history of IVC occurred in 1950, when

Wesley Whitten formulated a new IVC medium. The Krebs-Ringe bicabornate solution and BSA were added to this medium and led to an increase in the number of embryos successfully implanted. After Whitten, research sought to find out the nutritional needs of embryos and began to develop the microdrop culture technique (ANDREOTI, 2007).

Currently, the most widely used medium for handling and transporting oocytes to the laboratory is TCM 199 with a HEPES lid. This medium can be supplemented with amino acids, vitamins, nutrients and hormones (MAINARDES, 2010).

3.8 Techniques for assessing oocyte and embryo quality before and after CIVE

CIVE evaluation techniques are used to monitor the quality and activity of oocytes and embryos before and after culture.

3.8.1 Morphological and quantitative analysis

Morphological and quantitative analysis can be carried out using classical histology techniques. Histology morphologically analyzes the changes from sidewalk cells to cubic cells, with a single layer, called a unilaminar primary follicle. It also analyzes the structure of the oocyte during this follicular transition. The stages of histology are: fixation, dehydration, diaphanization, infiltration, inclusion, microtomy and staining of the slides (MATOS *et al.,* 2007).

3.8.2 Ultrastructural analysis by Transmission Electron Microscopy (TEM)

MET is the technique used to evaluate aspects of cytoplasmic organelles and ultrastructural changes in follicles, as well as to analyze the quality of oocytes and embryos used in IVC (MATOS *et al.,* 2007). The electron microscopes used for MET became commercially available in the 40s. Over the years, electron microscopy has developed and become increasingly sophisticated, becoming an independent area of research in which it has contributed to the knowledge of biochemical and physiological ultrastructure and the molecular biology of genes. MET transmits a beam of electrons through the sample, forming an image that is viewed through a phosphorescent screen, a film sensitizer or a digital camera (MOREIRA; LINS, 2010). It is extremely important to carry out ultrastructural analysis of the follicles after preservation techniques, such as cooling and cryopreservation, in order to verify the integrity of the oocyte or detect possible degenerative alterations (MATOS *et al.,* 2007).

3.8.3 Analysis of oocyte and embryo viability by fluorescence microscopy

The fluorescence emission is captured automatically with the help of specific fluorescence microscopy cameras, after exposure to the ultraviolet light of the microscope. The images generated are analyzed using camera image analysis software (GUENRA,

2013).

Morphology analysis is usually accompanied by oocyte and embryo viability analysis. The nucleus of the oocytes after IVM can be analyzed using fluorescence microscopy, which allows the chromosomes of the oocytes to be visualized and the stage of nuclear maturation to be determined. After IVM, the oocytes are stripped, marked with the nucleolus-specific dye Hoechst 33342 and then visualized using fluorescence inverted microscopy. This mechanism has been used to monitor the progression of IVM in bovine oocytes and other species (CHAVES, 2010).

Sometimes preservation techniques, such as cryopreservation, can cause the membrane of oocytes to rupture, which can lead to their death. Therefore, the use of dyes such as Trypan Blue is used to assess viability by analyzing the integrity of the cell membrane and cellular components (LUNARDI, 2011). These oocytes and embryos can then be analyzed using fluorescence microscopy, based on the simultaneous detection of live and dead cells marked by calcein acetoxymethyl (calcein-AM) and ethidium homodimer-1 (EthD-1), respectively (VAN DEN HURK *et al.,* 1998).

3.8.4 Analysis to detect apoptosis

Based on the morphology of apoptotic cells, there are techniques that analyze this cell death. One of the most widely used methods for verifying apoptosis in implanted oocytes and embryos is TUNEL (terminal deoxynucleotidyl transferase-mediated deoxyuridine triphosphate biotin nick end-labeling), which is essential for localizing nuclear DNA fragmentation. This localization is carried out by the enzyme terminal dioxygen transferase, which adds nucleotides to the DNA fragments. When observed, viable cells appear light-colored and those that have undergone apoptosis, dark-colored (SANTANA, 2009).

4 FINAL CONSIDERATIONS

Assisted reproduction biotechniques have emerged as an interesting possibility for the multiplication of animal reproduction. *In vitro* maturation systems offer favorable conditions for oocyte development, however, a better understanding of the particularities of physiology, substances, hormones and growth factors is needed to improve the efficiency of *in vitro* maturation of bovine oocytes.

It is known that oocyte maturation involves many particularities, from the amount/concentration of hormones and protein and energy sources to the manipulation of the oocytes during the development of this biotechnique. In this way, perfect harmony between these factors provides the oocyte competence necessary for optimal development *in vitro*.

PIVE makes it possible to increase the number of viable embryos from potentially fertile females of high genetic value and from animals that were not suitable for reproduction. The advantages of this technique have stimulated improvements in the stages, from follicular aspiration to embryo implantation. Improving the technique will guarantee the production of good quality embryos in considerable numbers and reduce problems caused by contamination from the handling of materials and equipment, or even human error. Such attitudes will contribute to a greater production of viable embryos and a reduction in the costs of PIVE, thus facilitating its management in Brazilian livestock.

5 REFERENCES

ADONA P. R.; LEAL C.L.V. Effect of concentration and exposure period to butyrolactone I on meiosis progression in bovine oocytes.
Arq.Bras.Med.Vet.Zootec., Belo Horizonte, v.58, n.3 p.354-359, 2006.

AGUILLAR J.J; WOODS G.L; MIRAGAYA M.H. Effect of homologous preovulatory follicular fluid on *in vitro* maturation of equine cumulusoocyte complexes. **Theriogenology**, v.56, n.5, p.745-758, 2001.

AKTAS, H. *et al.* Maintenance of bovine oocytes in prophase of meiosis I by high [cAMP]. **J. Reprod. Fertility and rep.**, v. 105, p. 227-235, 1995.

ANDRADE, E.R. *et al.* Consequences of production of reactive oxygen species in reproduction and main antioxidant mechanisms. **Rev.Bras.Reprod.Anim**., Belo Horizonte, v.34, n.2, p.79-85, 2010.

ANDRADE, G. A., *et al.* Factors affecting the pregnancy rate of recipients of bovine embryos produced *in vitro.* **Revista Brasileira de Reprodupao Animal**, Belo Horizonte, v.36, n.1, p.66-69, jan./mar. 2012.

ANDREOTI, M. ***In vitro* production of bovine embryos: use of glutathione during the sperm washing and capacitation process**. 2007. 45 f. Dissertation (Master's Degree in Animal Science) - Federal University of Para, Brazilian Agricultural Research Corporation - Eastérn Amazonia, Federal Rural University of Amazonia, Belem. 2007.

ANGUITA B. *et al.* Effect of oocyte diameter on meiotic competence, embryo development, p34 (cdc2) expression and MPF activity in prepubertal goat oocytes. **Theriogenology**, v.67, n.3, p.526-536, 2007.

BAGGIO, J. ***In vitro* production of bovine embryos**. 2008. 37 f. Monograph: (specialist in Bovine Production and Reproduction)- Castelo Branco University, Rio de Janeiro, 2008.

BILODEAU-GOESEELS, S. Cows are not mice: The role of cyclic AMP, phosphodiesterases, and adenosine monophosphate-activated protein kinase in the maintenance of meiotic arrest in bovine oocytes. **Mol. Reprod. Dev.**, v. 78, p.734-743, 2011.

BRAGANQA, J. F. M. **Estrus induction/synchronization hormonal strategies in beef heifers between 12 and 14 months of age**. 2007. 124 f. Thesis (Doctorate in Veterinary Medicine, area of concentration in Physiopathology of Reproduction)- Federal University of Santa Maria, Santa Maria, 2007.

BUENO, A. P.; BELTRAN, M. P. *In vitro* production of bovine embryos. **Revista Cientifica Eletronica de Medicina Veterinaria**, Garpa, n. 11, jul. 2008.

CAN A.; SEMIZ O.; CINAR O. Centrosome and microtubule dynamics during early stages of meiosis in mouse oocytes. **Mol.Hum.Reprod.**, v.9, n.12, p.749-756, 2003.

CARDOSO, D.; NOGUEIRA, G.P. Neuroendocrine mechanisms involved in heifer puberty. **Arq. Cienc. Vet. Zool**. Unipar, Umuarama, v. 10, n. 1, p. 59-67, 2007.

CARNEIRO G.F. *et al.* Influence of insulin-like growth factor-I and interaction with gonadotropins, estradiol, and fetal calf serum on *in vitro* maturation and parthenogenic development in equine oocytes. **Biol Reprod**, v.65, p.899-905, 2001.

CARNEIRO G.F. *et al.* Effect of IGF-I on the in vitro maturation of goat oocytes. *In:* REUNIAO ANUAL DA SOCIEDADE BRASILEIRA DE TECNOLOGIA DE EMBRIOES, 17, 2003, Beberibe, CE. **Proceedings.** Beberibe, CE: SBTE, p.286, 2003.

CHAVES, R. N., *et al. In vitro* culture systems for the development of mammalian immature oocytes. **Revista Brasileira de Reprodugao Animal**, Belo Horizonte, v.34, n.1, p37-49, jan/mar. 2010.

COTICCHIO G. *et al.* What criteria for the definition of oocyte quality? **Ann.N.Y.Acad.Sci**. V.1034. p. 132-144, 2004.

CROCOMO, L. F. *et al.* Peculiarities of oocyte collection for *in vitro* production of ovine embryos. **Revista Brasileira de Reprodugao Animal,** Belo Horizonte, v.36, n.1, p. 25-31, Jan/Mar. 2012.

CROSS, J.C. *et al.* Genes, development and evolution of the placenta. **Placenta**, v. 24, p. 123-130, 2003.

DAYAN, A.; WATANABE, M. R.; WATANABE, Y. F. Factors that interfere with the commercial production of IVF embryos. **Arquivos da Faculdade de Veterinaria da UFRGS**, Porto Alegre, v. 27, n. 1, p. 181185, 2000.

DODE, M. A. N. *et al.* The effect of sperm preparation and time of coincubation on *in vitro* fertilization of Bos indicus oocytes. **Animal Reproduction Science**, v. 69, n.1-2, p. 15- 23, 2002.

DOMINGUES, M. C. N., *et al.* Viability of vitrified embryos from *in vitro* fertilization of oocytes from cows supplemented with canola. **Arquivo Brasileiro de Medicina Veterinaria e Zootecnia**, Maringa, v.66, n.1, p.145-151, 2014.

DOMINGUES, S. F. S., *et al.* Breeding biotechnologies as a complementary strategy to *in situ* conservation of endangered neotropical primates: perspectives and challenges. **Revista Brasileira de Reprodupao Animal**, Belo Horizonte, v.35, n.2, p124-129, Apr/Jun. 2011.

ELSAYED, A. *et al.* Large-scale transcriptional analysis of bovine embryo biopsies in relation to pregnancy success after transfer to recipients. **Physiol Genomics,** v. 28, p.84-96, 2006.

FAIR, T. *et al.* Global gene expression analysis during bovine oocyte *in vitro* maturation. **Theriogenology**, v.68, n.1, p.91-97, 2007.

FERNANDES C.B. ***In vitro* maturation of equine oocytes: comparison between TCM 199, SOFaa and HTF: BME media, and evaluation of the addition of bovine FSH, equine FSH and equine growth hormone by means of oocyte transfer.** 2004. 129f. Dissertation (Master's Degree in Veterinary Medicine), Post-graduation in Veterinary Medicine, Universidade Estadual Paulista, 2004.

FERREIRA E. *et al.* Cytoplasmic maturation of bovine oocytes: Structural and biochemical modifications and acquisition of developmental competence. **Theriogenology**, v.71, n.5, p. 836-848, 2008.

FERREIRA, M. A. **Nuclear and cytoplasmic maturation of bitch oocytes collected at different stages of the estrous cycle and cultured *in vitro* in sequential media with hormones and sperm**. 2010. 95f. Thesis (Doctorate in Veterinary Medicine -

Animal Reproduction)- Universidade Estadual Paulista, Jaboticabal. 2010.

GANDOLFI, B.T.A.L.; GANDOLFI, F. The maternal legacy to the embryo: cytoplasmic components and their effects on early development. **Theriogenology**, v.55, n.6, p.1255-1276, 2001.

GARCIA, J. M.; AVELINO, K. B.; VANTINI, R. State of the art of *in vitro* fertilization *in* cattle. In: SIMPOSIO INTERNACIONAL DE REPRODUQAO ANIMAL APLICADA, 1., 2004, Londrina. **Proceedings...** Londrina: UEL, 2004. p. 223-230.

GUEMRA, S.; MONZANI, P.S.; SANTOS, E.S.; ZANIN, R.; OHASHI, O.M.; MIRANDA, M.S.; ADONA, P.R. *In vitro* maturation of bovine oocytes *in* media supplemented with quercetin and its effect on embryonic development. **Arq. Bras. Med. Vet. Zootec.**, v.65, n.6, p.1616-1624, 2013.

GILCHIST, R.B.; THOMPSON, J. G. Oocyte maturation: emerging concepts and technologies to improve developmental potential *in vitro*. **Theriogenology**, v.67, n.1, p. 6- 15, 2007.

GONQALVES, P. B. D. *et al. In vitro* embryo production. In: GONQALVES, P. B. D.; FIGUEIREDO, J. R.; FREITAS, V. J. F. **Biotechniques applied to animal reproduction**. Sao Paulo: Varela, p. 195-226, 2002.

GONQALVES P.B.D. *et al. In vitro* production of bovine embryos: the state of the art**. Rev Bras.Reprod.Anim**., Belo Horizonte. v.31, n.2 p.212-217, 2007.

GONQALVES P.B.D. *et al.* ***In vitro* embryo production. In: Biotechniques applied to animal reproduction.** 2.ed. Sao Paulo: Roca, 2008. p.261-291.

GOTTARDI F.P.; MINGOTI G.Z**.** Bovine oocyte maturation and influence on subsequent embryonic developmental competence**. Rev.Bras.Reprod.Anim.**, Belo Horizonte, v.33, n.2, p.82-94, 2009.

GOTTARDI, F. P; MINGOTI, G. Z. Bovine oocyte maturation and influence on the acquisition of competence for embryo development. **Revista Brasileira de Reprodupao Animal**, Belo Horizonte, v.34, v.2, p. 82-94, Apr/Jun. 2009.

GOUVEIA, F. F. ***In vitro* production of bovine embryos**. 2011. 35 f. Monograph (degree in Veterinary Medicine) - Veterinary Medicine Course, University of Brasilia. Brasilia, 2011.

GUEMRA, S., *et al. In vitro* maturation of bovine oocytes in media supplemented with quercetin and its effect on embryonic development. **Arquivo Brasileiro de Medicina Veterinaria e Zootecnia**, Tamarana, v.65, n.6, p.1616-1624, 2013.

GUIXUE Z.; LUCIANO A.M.; COENEN K. The influence of cAMP before or during bovine oocyte maturation on embryonic developmental competence. **Theriogenology**, v. 55, n.8, p. 1733-1743, 2001.

HAMATANI, T. *et al.* What can we learn from gene expression profiling of mouse oocytes? **Reproduction**, v.135, n.5, p.581-592, 2008.

HUMBLOT, P. *et al.* Effect of stage of follicula growth during superovulation on developmental competence of bovine oocytes. **Theriogenelogy**. v.63, p. 1149-1166, 2005.

JONES, K.T. Turning it on and off: M-phase promoting factor during meiotic maturation and fertilization. **Mol. Hum. Reprod**., v.10, p.1-5, 2004.

KATZ-JAFFE *et al.* Transcriptome analysis of in vivo and *in vitro* matured bovine MII oocytes. **Theriogenology**, v.71, n.6, p.939-946, 2009.

KUBELKA, M. *et al.* Butyrolactone I reversibly inhibits meiotic maturation of bovine oocytes, without influencing chromosome condensation activity. **Biology of Reproduction**, v.62, n.2, p.292-302, 2000.

KRISHER R.L. The effect of oocyte quality on development. **J.Anim.Sci.**, v.82, n.13, p.45-51, 2004.

LOIOLA, M. V. G. **Validation of an *in vitro* bovine embryo production program with oocyte and embryo transport over long distances**. 2013. 61 f. Dissertation (Master's Degree in Tropical Animal Science)- Federal University of Bahia, Salvador. 2013.

LONERGAN P.; FAIR T. *In vitro-produced* bovine embryos- Dealing with the warts. **Theriogenology**, v.69, n.1, p.17-22, 2008.

LOPES, C. A. P. **Use of the biotechnique of manipulation of oocytes included in pre-antral follicles for the *in vitro* evaluation of the potential of anti-zona pellucida antibodies for the immuno-sterilization of bitches**. 2008. 154 f. Thesis (Doctorate in Veterinary Medicine, area of concentration Reproduction and Animal Health) - State University of Ceara, Fortaleza, 2008.

LUCIANO, A. M. *et al.* Effect of different cryopreservation protocols on cytoskeleton and gap junction mediated communication integrity in feline germinal vesicle stage oocytes. **Cryobiogy**. V. 59, n.1, p. 90-105, 2009.

LUNARDI, F. O. **Analysis of the morphology, viability and ultrastructure of preantral follicles after vitrification of ovine ovarian tissue**.
2011. 96 f. Dissertation (Master's Degree in Veterinary Sciences, area of concentration Reproduction and Animal Health). - State University of Ceara, Fortaleza. 2011.

MAINARDES, G. A. **Follicular aspiration, *in vitro* production**

and manipulation of bovine embryos. 2010. 93 f. Monograph (Graduation in Veterinary Medicine) - Veterinary Medicine Course - Universidade Tuiuti do Parana, Curitiba, 2010.

MAMO, S. *et al.* Sequential analysis of global gene expression profiles in immature and *in vitro* matured bovine oocytes: potential molecular markers of oocyte maturation. **BMC Genomics**, v. 12, n.151, p.1- 14, 2011.

MATOS, M. H. T.; *et al.* Techniques for assessing the quality of pre-antral ovarian follicles cultured *in vitro.* **Revista Brasileira de Reprodupao Animal**, Belo Horizonte, v.31, n.4, p.433-442, Oct/Dec. 2007.

MAYES M.; SIRARD M.A. The influence of cumulus-oocyte complex morphology and meiotic inhibitors on the kinetics of nuclear maturation in cattle. **Theriogenology**, v.55, n.4, p.911-922, 2001.

MEINECKE, B. *etal.* Histone H1 and MAP kinase activities in bovine oocytes following protein synthesis inhibition. **Reproduction in Domestic Animals**, v. 36, p.183-188, 2001.

MERTON, S.S. *et al.* Factors affecting oocyte quality and quantity in commercial application of embyos technologies in the cattle beeding industry. **Theriogenology**, v. 59, n. 2, p. 651-674, 2003.

MINGOTI, G.Z. **Oocyte maturation associated with steroidogenesis.**
Role of blood serum, serum albumin and steroid hormones. Sao Paulo: Faculty of Medical Sciences, University of Sao Paulo Paulo, 2000. 135p. Dissertation (Doctorate in Physiology) - Faculty of Medical Sciences, University of Sao Paulo, 2000.

MINGOTI G.Z. **Technical aspects of *in vitro* production of bovine embryos. In: Advanced Topics in Reproductive Biotechnology**, 2005, Jaboticabal, SP. Jaboticabal, SP: Funep, 2005.

MONTAGNER, M. M., *et al.* Hepes in Embryo Production *in vitro.* **Ciencia Rural**, Santa Maria, v.30, n. 3, p. 469-474, 2000.

MOREIRA, D. S. M; LINS, U. G. C. Microscopy techniques for quantifying and identifying microorganisms. **Saude & Ambiente em Revista**. Duque de Caxias, v.5, n.2, p.01-11, jul/dez 2010.

PARKS, J.C. *et al.* Blastocyst gene expression correlates with implantation potential. **Fertil Steril**, v. 95, p. 1367-1372, 2011.

PONTES J.H.F *et al.* Comparison of embryo yield and pregnancy rate between *in vivo* and *in vitro* methods in the same Nelore (Bos indicus) donor cows. **Theriogenology**, v. 71, n.4, p.690-697, 2009.

PONTES, J.H.F *et al.* Ovum pick up, *in vitro* embryo production, and pregnancy rates from a large-scale commercial program using Nelore cattle (Bos indicus) donors. **Theriogenology**, v. 75, n. 9, p. 1640-1646, 2011.

RADOSTITS, O. M. HERD HEALTH **Food Animal Production Medicine**. 3ª ed. Philadelphia: W. B. Saunders Company, 2001.

RUBIN, M. I. B. *et al. In vitro* embryo production and cloning: a known path? **Revista Brasileira de Reprodupao Animal**, Belo Horizonte, v.36, n.6, p. 77-85, dec. 2009.

SANDRI, LR. **The effect of meiotic blockers on the maturation and ultrastructure of oocytes and their effect on embryo production *in vitro*.** 2007. 58f. Dissertation (Master's degree in veterinary medicine) - Federal University of Santa Maria, Rio Grande do Sul, 2007.

SANTANA, P. P. B. **Apoptosis in bovine embryos matured and cultured *in vitro* with dexamethasone**. 2009. 65 f. Monograph (graduation in Biomedicine)- Biomedicine Course, Federal University of Para. Belem, 2009.

SANTOS, K.J.G. **Effect of progesterone on embryo production in Gir and Girolando heifers:** effect of exogenous progesterone on embryo production in dairy heifers. 2010. 114f. Thesis (Doctorate in animal science) - Universidade Federal de Goias,

Goiania, 2010.

SIRARD M.A. Resumption of meiosis: mechanism involved in meiotic progression and its relation with developmental competence.
Theriogenology, v.55, n.6, p.1241-1254, 2001.

SIRARD M.A, *et al.* Contribution of the oocyte to embryo quality. **Theriogenology**, v.65, n.1, p.126-136, 2006.

SOUZA, L. C. **Embryo transfer and *in vitro* fertilization (IVF) in cattle**. Monograph (specialist in Bovine Production and Reproduction) - Castelo Branco University, Rio de Janeiro, 2007.

STOJKOVIC, M.*et al.* Mitochondrial distribution and adenosine triphosphate content of bovine oocytes before and after *in vitro* maturation: correlation with morphological criteria and developmental capacity after *in vitro* fertilization and culture. **Biology of Reproduction**, v.64, p.904-909, 2001.

TAKAGI, M.; *et al.* Impaired final follicular maturation in heifers after superovulation with recombinant thuman FSH. **Reproduction**, Cambridge, v. 121, n. 6, p. 941-951,2001.

TILLY, J. L.; JOHNSON, J. Recent arguments against germ cell renewal in the adult human ovary. **Cell cycle**, v. 8, p. 879-883, 2007.

TSAFRIRI, A. *et al.* Oocyte maturation involves compartmentalization and opposing changes of cAMP levels in follicular somatic and germ cells: studies using selective phosphodiesterase inhibitors. **Dev. Biol**., v. 178, p. 393-402, 1996.

ULLAH, Z. *et al.* Differentiation of trophoblast stem cells into giant cells is triggered by p57/Kip2 inhibition of CDK1 activity. **Genes**, v. 22, p. 30243036, 2008.

ULLOA, S. M. B., *et al.* Effects of different oocyte retrieval and *in vitro* maturation systems on bovine embryo development and quality. **Zygote,** Mariensee, v. 23, p. 367-377, jun. 2014

VAN DER HURK R; ZHAO J. Formation of mammalian oocytes and their growth, differentiation and maturation within ovarian follicles.
Theriogenology, v.63, n.6, p. 1717-1751, 2005.

VARAGO, F. C.; MENDONQA, L. F.; LAGARES, M. A. *In vitro* production of bovine embryos: state of the art and perspective of a technique in constant evolution. **Revista Brasileira de Reprodupao Animal**, v. 32, n. 2, p. 100-109, 2008.

WANG, W.; DAY, B.N.; WU, G. How does polyspermy happen in mammalian oocytes? **Microscopy Research and Technique**, v.61, p.335-341,2003.

WASSARMAN, P. M.; ALBERTINI, D. F. The mammalian ovum. In:
Knobil, E.; Neil, J. D. **The physiology of Reproduction**, New York: Raven, NY, p.9-122,1994.

Printed by Books on Demand GmbH, Norderstedt / Germany